Formation Mechanism on Fullerene Nanoparticles

Sylvester Gyasi

Bibliographic information published by the German National Library:

The German National Library lists this publication in the National Bibliography; detailed bibliographic data are available on the Internet at http://dnb.dnb.de.

ISBN: 9783346865441
This book is also available as an ebook.

Contents

List of Figures

List of Tables

Introduction

Table 1: Description of similarities and differences between carbon nanomaterials. Adapted from Kroto et al., 1985; Dresselhaus et al 2010; Fang et al., 2020; Hass et al., 2008; Balandin et al., 2008 and Enoki et al., 2003.

Description	Graphene	Fullerene	Carbon Nanotubes
Discovery	Demonstrated by Andre Geim and Konstantin Novoselov from the university of manchester in 2004.	Eiji Osawa predicted the existence of C_{60} in 1970 and was observed in 1985 by Richard E. Smalley, Robert F. Curl and Harold W. Kroto. Macroscopic synthesis is achieved in 1990.	Identified by Sumio Lijima in 1991.
Definition	An allotrope of carbon that occurs as sheets of carbon.	Allotrope of carbon that occurs as spheres of carbon.	Allotrope of carbon that occurs as cylindrical of carbon.
Structure	2 dimensional	Zero dimensional	1 dimensional
Synthesis	Mainly chemical vapor deposition	Mainly combustion and Arc discharge	Mainly chemical vapor deposition
Hybridization	Sp^2 Hybridized	Sp^2 Hybridized	Sp^2 Hybridized
Edges	There are edges.	There are no edges because it is a closed structure.	There are edges.
Appearance	Sheets	Spheres or ellipsoid tubes.	Cylindrical
Chemical Bonds	Single bonds between carbon atoms.	Single and double bonds between carbon atoms.	Two bonds are single, and one is double.
Pi electron clouds	Contains pi electron clouds.	No pi electron clouds.	Contains pi electron clouds.
Electrical conductivity	Can conduct electricity due to delocalised electrons.	Cannot conduct electricity because delocalised electrons cannot move.	Can conduct electricity due to delocalised electrons.

Melting point	High melting point due to strong covalent bonds	Lower melting point due to weak intermolecular forces	High melting point due to strong covalent bonds
Types	Graphene Graphene Oxide Reduced Graphene Oxide.	Higher number fullerene and fullerene derivatives.	Single walled nanotubes Double walled nanotubes Multi Walled nanotubes.
Conductivity	Highly transparent and conductive	Semiconducting	Can be metallic and semiconducting
Thermal conductivity	Up to 5300W/m.K	0.4 W/m.K	Up to 3000W/m.K
Electron Mobility	>15000cm^2 / V.s	$1.5 \times 10 -2$ cm^2 / V.s	200–16000 cm^2 / V.s
Resistivity	10^{-6} Ω cm	10^{-6} Ω cm	10^{-6} Ω cm
Young modulus	~1100 GPa	53-69 GPa	Over 1 TPa
Industrial Applications	used in electronics, energy storage, sensors, coatings.	carrier for gene and drug delivery systems.	used to make aircraft and spacecraft bodies

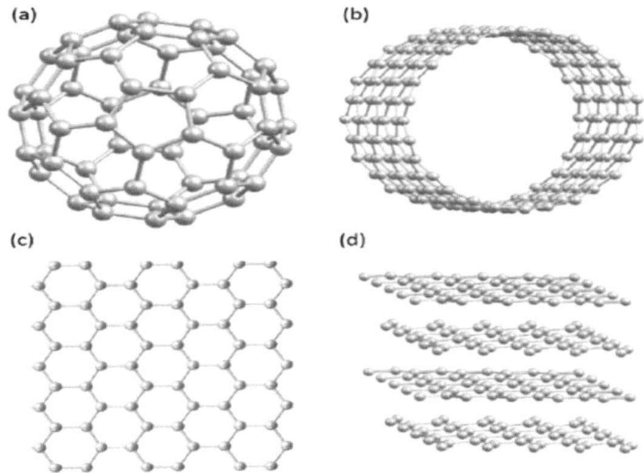

Figure 1: (a) Zero- dimensional Fullerene, (b) one- dimensional Carbon Nanotube, (c) two-dimensional Graphene and (d) three- dimensional Graphite. Adapted from Santhiran et al., (2021).

Formation mechanism

The exact pathway of these mechanism can depend on variety of factors including the method of synthesis, the reaction conditions such as the temperature, pressure, presence of catalysts and the starting materials used. The formation mechanism of fullerene involves the following steps: Nucleation, Growth, Rearrangement, Fusion, Annealing and purification (Curl et al., 1993).

Proposed Formation Mechanism

The proposed formation mechanism of fullerenes is based on a combination of experimental observations and theoretical models. In the early 1990s, researchers discovered that carbon clusters produced by methods such as laser ablation or arc discharge contained closed-cage structures consisting of pentagonal and hexagonal rings, which were later identified as fullerenes. The proposed formation mechanisms include (Goeres et al., 1991):

Special Pathways Based on Particular Intermediate Size Clusters Mechanism.

The formation mechanism of fullerenes can involve special pathways based on intermediate size clusters, which can lead to the formation of fullerenes with specific sizes and structures. One example of a special pathway involves the use of intermediate-sized carbon clusters called carbon onions. Carbon onions are made up of concentric shells of carbon atoms, like the layers of an onion. Researchers have found that by using carbon onions as starting materials, they can selectively form fullerenes with a specific number of carbon atoms, such as C_{60} or C_{70}. This is because the onion structure provides a template for the formation of a closed-cage fullerene, with the number of concentric shells determining the size of the final structure (Raghavachari et al., 1992).

Another example of a special pathway involves the use of endohedral fullerenes, which are fullerenes that contain an atom or molecule inside the cage. Endohedral fullerenes can be formed by incorporating the desired guest molecule into the starting material, which can then undergo the normal pathways of fullerene formation. The presence of the guest molecule can influence the structure and stability of the final fullerene product, as well as its potential applications in areas such as drug delivery or energy storage (Strout et al., 1993).

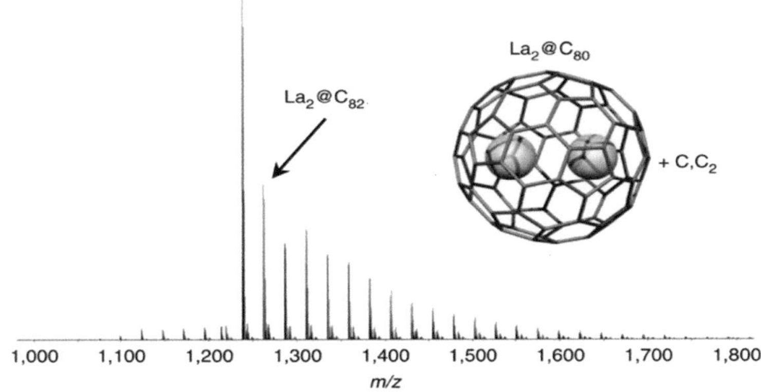

Figure 2: Larger La$_2$@C$_{2n}$ endohedral fullerenes formed by incorporating guest molecule. Adapted from Mercado et al., 2009.

The Pentagon Road Mechanism

The pentagon road is an appealing mechanism which begins with the formation of a small carbon cluster, which may contain one or more pentagonal rings. As the cluster grows, new carbon atoms are added to the edges of the existing rings, eventually forming a curved structure resembling a tube or sphere. However, in order to close the structure and form a fullerene, additional pentagonal rings must be added to the growing cluster (Curl et al., 1994).

The pentagon road mechanism proposes that these pentagonal rings are added in a stepwise fashion, with each new ring being formed by the rearrangement and fusion of neighbouring hexagonal rings. Specifically, a pair of adjacent hexagonal rings are transformed into a pentagon and a heptagon through the addition of two carbon atoms, followed by the rearrangement of the neighbouring hexagons to accommodate the new rings. This process can be repeated multiple times, creating a zigzagging pathway of pentagonal rings that ultimately leads to the closure of the fullerene structure (Smalley et al., 1992).

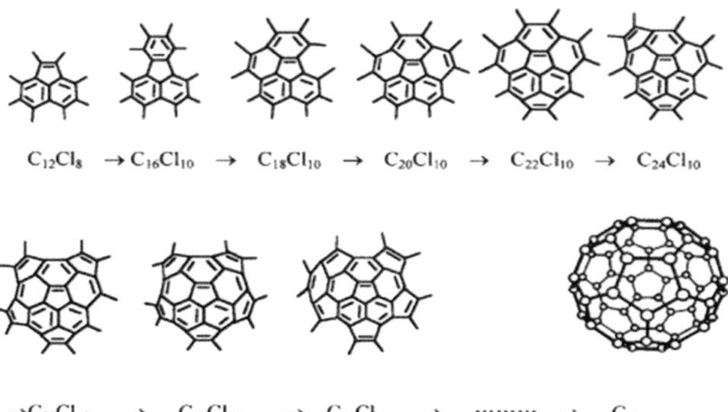

Figure 3: An image of the growth mechanism of C_{12} to C_{60}. Adapted from Xie et al., 2000.

The Fullerene Road Mechanism

The fullerene road involves the growth and fusion of carbon clusters into closed-cage structures. This mechanism is thought to be responsible for the formation of the most common fullerene species, including C_{60} and C_{70} (Heath, 1991). The fullerene road begins with the nucleation of small carbon clusters, which can be initiated by a variety of methods such as laser ablation, arc discharge, or chemical vapor deposition. These clusters then grow through the addition of new carbon atoms, which may be incorporated one-by-one or in small groups. As the cluster grows, it undergoes a series of structural rearrangements, including the formation of pentagonal and heptagonal rings, which can introduce strain into the structure. The fullerene road proposes that the cluster continues to grow and rearrange until it reaches a critical size, at which point it undergoes a process of closure to form a closed-cage fullerene. The exact mechanism of closure is not fully understood, but it may involve the fusion of neighbouring carbon clusters, the collapse of a carbon nanotube, or other processes (Smalley et al., 1991).

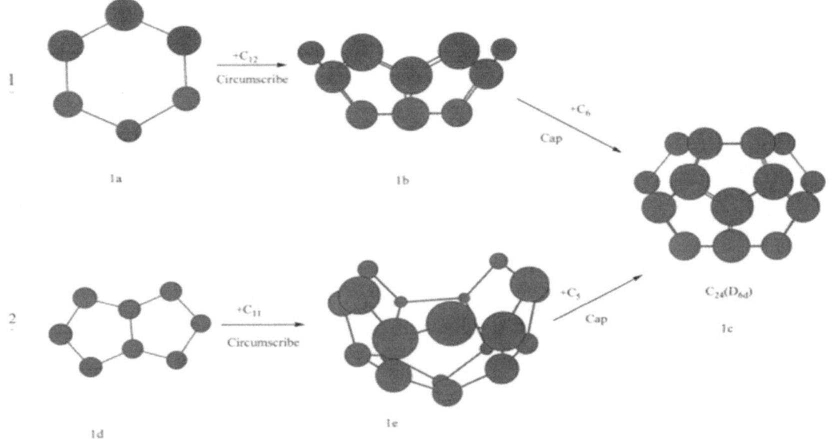

Figure 4: Formation of C_{24} (D_{6d}) from precursor 1a and 1b. Adapted from Lin et al., 2003.

Ring Stacking Mechanism

This mechanism involves the stacking of smaller carbon rings to form larger ones, ultimately leading to the formation of fullerene cages. These rings can form from larger carbon clusters or from carbon precursors such as hydrocarbons. The small carbon rings then stack together to form larger rings. This can occur through several mechanisms, such as the addition of new rings to the edge of an existing ring or the fusion of two or more smaller rings together. As the rings continue to stack and fuse, they eventually form a closed cage-like structure like a fullerene. The resulting structure can have various sizes and shapes, depending on the number and arrangement of the carbon rings (Curl et al., 1993).

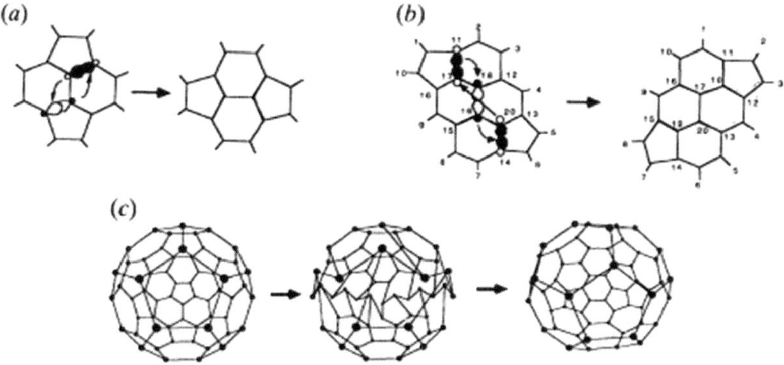

Figure 5: Ring arrangement process of C_{60}. Adapted from Curl et al., 1993.

Experimental Evidence

To comprehend how fullerenes are formed during their synthesis, researchers attempted to simulate the most efficient growth pathway using a semi-empirical quantum mechanics program. The model focused on C_2 addition, which results in the formation of C_{60}, and examined three primary pathways: cyclic ring, pentagon, and fullerene road (Hutter et al., 1994).

Molecular Modelling Method

For this computational model, a semi-empirical quantum mechanical approach was chosen due to its ability to accurately model nano clusters in less time. Three semi-empirical methods were employed: Modified Neglect of Differential Overlap (MNDO), Austin Model 1 (AMI), and Parameterized Model 3 (PM3). Since the electronic structures of fullerenes differ from other hydrocarbon structures, empirical methods were necessary. The MNDO method produced the most precise results in this experiment (Stewart et al., 1989).

Justification for C_2 Insertion-Based Growth.

The growth of C_{60} can be simulated through the step-by-step addition of carbon molecules, such as C_1, C_2, C_3, or larger. However, observations from regenerative sooting discharge's emission spectra suggest that C_2 is a significant component of carbonaceous plasma. This presence of C_2 may play a crucial role in the formation of fullerenes (Ahmad et al., 2000).

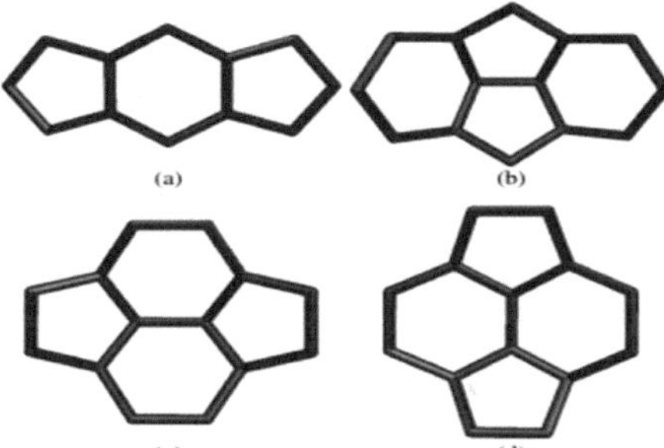

Figure 6: An image of (a) two pentagons and a central hexagon before C_2 insertion, (b) C_2 insertion leading to the addition of a new hexagon, (c) pyracylene/pyracene patch which can undergo SW transformation, (d) pyracylene/pyracene patch after SW transformation. Adapted from Khan et al., (2006)

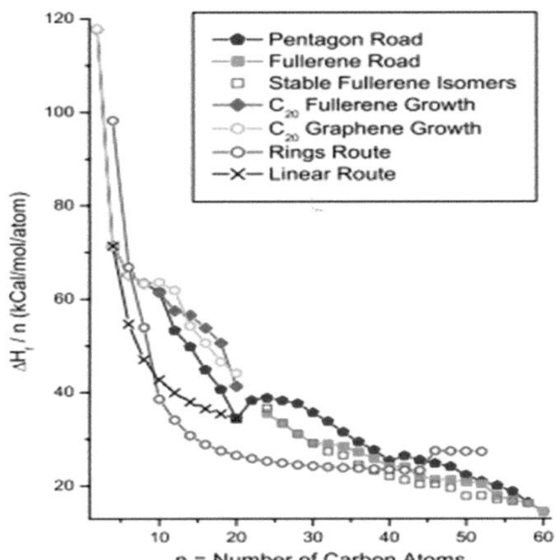

Figure 7: Growth of C_2 into C_{60} by C_2 insertion, for n = 2–20 energies of C_{20} fullerene, graphene, corannulene, linear chain and ring growth. Adapted from Khan et al., 2006.

In the case of fullerenes that have identical atoms, the ΔH_f per carbon atom is almost equal to the negative value of the binding energy per atom, which is the energy required to extract an atom from the molecule. The stability of a molecule increases with the strength of atom binding, so the ΔH_f per atom can be used as a metric for assessing molecular stability. The molecule is more stable if the value of ΔH_f per atom is lower (Khan et al., 2006).

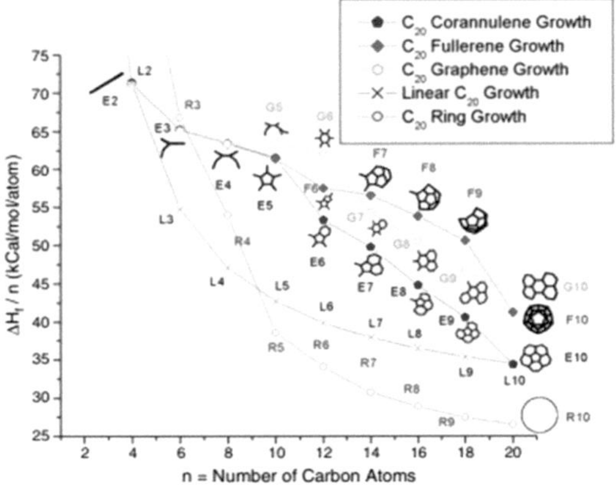

Figure 8: An image of the growth of C_{20} corannulene, fullerene, graphene, linear chain and ring is modelled from C_4 up to C_{20}. Adapted from Khan et al., 2006.

The provided figure displays the growth sequence of four potential C_{20} isomers. When individual carbon dimers merge to create bigger carbon clusters, they tend to arrange themselves in linear chains, which is quite noticeable. As these linear chains continue to grow, additional ring structures become increasingly energetically favourable.

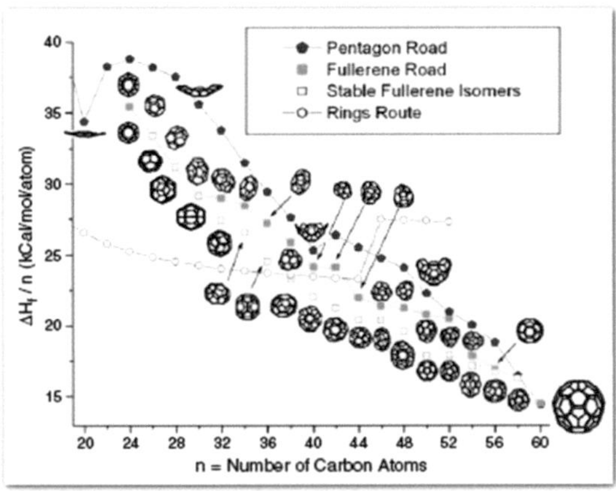

Figure 9: The graph of growth of C_{60} from C_{24} modelled by both pentagon and fullerene roads. Addition of the growth of C_{60} ring is also shown. Adapted from Khan et al., (2006)

The provided figure demonstrates that, from an energetic perspective, growth through the fullerene route is preferable than growth through the pentagon route. Geometry-optimized structures of these closed-cage systems are also depicted. It's evident that the pentagon route is less energetically favourable than the fullerene route, which may be due to the high strain present in the edge atoms of open-cage fragments (the pentagon route).

Conclusion on the Experiment

The initial growth begins with linear carbon chains and, when n = 10, the chains start to form rings. These rings continue to grow by ingesting C_2 molecules. Around n = 38, the rings appear to transition into fullerenes, and the growth of C_{60} proceeds through the fullerene route. According to symmetry considerations, the transition from ring to fullerene may occur much earlier than n = 38 in a collision-dominated environment during the synthesis of fullerenes (Khan et al., 2006).

Differences Between the Proposed Mechanisms

Table 2: Differences between the proposed mechanism Adapted from Kroto et al., 1985; Dresselhaus et al 2010; Fang et al., 2020; Hass et al., 2008.

Description	Pentagon road	Fullerene road	Ring stacking
Role of pentagonal rings in final fullerene structure.	proposes that pentagonal rings are added in a stepwise fashion to a growing carbon cluster, leading to the formation of a zigzagging pathway of pentagonal rings.	the pentagonal rings are formed as part of the growth and rearrangement of the carbon cluster, without the need for a specific pathway.	plays a role in determining the final fullerene structure, as they affect the curvature and shape of the molecule.
Emphasis on kinetic versus thermodynamic factors.	proposes a more deterministic pathway for fullerene formation, in which the stepwise addition of pentagonal rings leads to a predictable structure.	places greater emphasis on kinetic effects, suggesting that the specific conditions of the reaction can lead to the formation of certain fullerene structures even if they are not the most thermodynamically stable.	Kinetic factors influence the size and shape of the fullerene molecules formed. thermodynamic factors can influence the stability and yield

			of the final product
Roles of strain	emphasizes the importance of introducing strain into the growing carbon cluster through the addition of pentagonal rings	proposes that the cluster undergoes a series of structural rearrangements, which may or may not involve the formation of pentagonal rings, to reach a closed-cage fullerene structure.	The isolated pentagon rule states that a fullerene molecule with no adjacent pentagonal rings will have the lowest possible strain energy.
Closure mechanism	suggests that closure occurs through the stepwise addition of pentagonal rings, with closure occurring when the structure reaches a certain size.	proposes that closure occurs through a series of thermally activated rearrangements, which may involve the fusion of neighbouring carbon clusters or the collapse of a carbon nanotube	two half-fullerenes come into contact, forming a transition state that involves a pentagon and a heptagon sharing an edge
Dependence on reaction conditions	proposes a more deterministic pathway, with the stepwise addition of pentagonal rings leading to a predictable fullerene structure.	places greater emphasis on the importance of reaction conditions in determining the final fullerene structure.	Places much importance on reaction conditions as it can affect the yield, size, and properties of the resulting fullerenes
Predictive power	used to predict the structures of a wide range of fullerenes, including those with	used to predict the formation of unusual or unexpected fullerene structures, such as the	to predict the stability and structure of fullerenes, specifically the

| | non-integer numbers of carbon atoms. | "bucky onion" or "bucky egg" structures. | formation of one-dimensional (1D) stacks of fullerenes |

Outlook

Table 3: Bottleneck and mitigation. Adapted from Kroto et al., 1985; Dresselhaus et al 2010; Fang et al., 2020; Hass et al., 2008; Balandin et al., 2008 and Enoki et al., 2003.

Bottleneck/Problem	Description	Mitigation/Solution
Cost	Carbon nanomaterials can be expensive to produce in large quantities, especially those with high levels of purity and uniformity.	Improving production efficiency, Recycling and reusing, Collaboration and funding.
Toxicity	Some types of carbon nanomaterials, such as carbon nanotubes, can potentially cause harm to human health if they are inhaled or ingested.	More research is needed to fully understand the potential risks and develop safe handling procedures.
Scalability	Scaling up the production of carbon nanomaterials while maintaining their quality and properties can be difficult.	optimizing the production process. This includes improving the efficiency of the process, reducing waste, and increasing throughput.
Structural Complexity	Carbon nanomaterials come in a wide range of structures and sizes, which can make it challenging to control their properties and tailor them for specific applications.	Better characterization techniques can help researchers understand the properties and behaviour of carbon nanomaterials more accurately. This can involve developing new microscopy and spectroscopy methods, as

		well as improving existing techniques.

Market Penetration

Table 4 Market penetration factors. Adapted from Kroto et al., 1985; Dresselhaus et al 2010; Fang et al., 2020; Hass et al., 2008; Balandin et al., 2008 and Enoki et al., 2003.

Factor	Description
Performance	Carbon nanomaterials have unique properties that can offer superior performance in various applications compared to traditional materials. Therefore, demonstrating superior performance in relevant applications can help to drive market adoption.
Cost	Developing cost-effective production methods and scaling up production to achieve economies of scale are critical to reducing the cost of these materials.
Regulations	Ensuring compliance with relevant regulations and demonstrating the safety of these materials are crucial to gaining market acceptance.
Sustainability	Carbon nanomaterials have the potential to offer sustainability benefits, such as reduced energy consumption and waste production, and highlighting these benefits could help to drive market adoption.
Awareness	Educating potential customers about the unique properties and potential applications of these materials can help to drive demand and promote wider adoption.

References

Stephens, P. W.; Xu, C.; Furenlid, L. R.; McDonald, R. N.; Wagner, J. P.; Krishnan, R. S.; Chouteau, G. et al. J. (1991) Phys. Chem., 95, 7564-7568.

Sabih D Khan and Shoaib Ahmad (2006) Modelling of C2 addition route to the formation of C_{60} Nanotechnology 17 4654. DOI 10.1088/0957-4484/17/18/021

Goeres, A. and Sedlmayr, E. (1991) On the nucleation mechanism of effective fullerite condensation. Chem. Phys. Lett. 184, 310.

Raghavachari, K. & Rohlfing, C. M. (1992). Imperfect fullerene structures: isomers of C60. Chem. Phys. Lett. 96, 2463-2466.

Hutter J, Luthi H P and Diederich F (1994) J. Am. Chem. Soc.166 750–6.

Mercado, B. Q., Haywood, J., Wang, Y., Duncan, M. A., & Khanna, S. N. (2009). Isolation and structural characterization of the molecular nanocapsule Sm2@D3d (822)-C104. Angewandte Chemie International Edition, 48(48), 9114-9116.

Stewart, J. J. P. (1989). Optimization of parameters for semiempirical methods V: Modification of NDDO approximations and application to 70 elements. Journal of Computational Chemistry, 10(2), 209-220.

Ahmad S, Qayyum A, Akhtar M N and Riffat T (2000) Nucl. Instrum Methods Phys. Res. B 171 551–7.

Khan, Sabih & Ahmad, Shoaib. (2006). Modelling of C-2 addition route to the formation of C-60. Nanotechnology. 17. 4654-8. 10.1088/0957-4484/17/18/021.

B.G Demczyk, Y.M Wang, J Cumings, M Hetman, W Han, A Zettl, R.O Ritchie, (2002) Direct mechanical measurement of the tensile strength and elastic modulus of multiwalled carbon nanotubes, Materials Science and Engineering: A, Volume 334, Issues 1–2, Pages 173-178, ISSN 0921-5093,

Xiao-Lin Xie, Yiu-Wing Mai, Xing-Ping Zhou, (2005) Dispersion and alignment of carbon nanotubes in polymer matrix: A review, Materials Science and Engineering: R: Reports, Volume 49, Issue 4, Pages 89-112, ISSN 0927-796X

R. Nagarajan (2008), "Nanoparticles: Building Blocks for Nanotechnology," in Nanoparticles: Synthesis, Stabilization, Passivation, and Functionalization, American Chemical Society, pp. 2-14.

Iijima, S. (1991). Helical microtubules of graphitic carbon. Nature, 354(6348), 56-58. doi: 10.1038/354056a0.

Stankovich, S., Dikin, D. A., Piner, R. D., Kohlhaas, K. A., Kleinhammes, A., Jia, Y., Wu, Y., Nguyen, S., and Ruoff, R. S. (2007). Synthesis of graphene-based nanosheets via chemical reduction of exfoliated graphite oxide. Carbon, 45, 1558-1565. doi: 10.1016/j.carbon.2007.02.034

Zhang, R., Dong, Y., Kong, W., Han, W., Tan, P.-H., Liao, Z.-M., Wu, X., & Yu, D. (2012). Growth of large domain epitaxial graphene on the C-face of SiC. Journal of Applied Physics, 112(10), 104307. doi: 10.1063/1.4766405.

H. W. Kroto, J. R. Heath, S. C. O'Brien, R. F. Curl, and R. E. Smalley (1985), "C60: Buckminsterfullerene," Nature, vol. 318, pp. 162–163. doi: 10.1038/318162a0

Dresselhaus, M. S., Dresselhaus, G., & Saito, R. (2010). Carbon Nanomaterials: Synthesis, Structure, Properties, and Applications. Springer.

Enoki, T.; Endo, M.; Suzuki (2003), M. Graphite Intercalation Compounds and Applications; Oxford University Press: New York, NY, USA; pp. 452.

Santhiran, A., Iyngaran, P., Abiman, P., & Kuganathan, N. (2021). Graphene Synthesis and Its Recent Advances in Applications—A Review. C, 7(4), 76. doi: 10.3390/c7040076

Fang, B., Chang, D., Xu, Z., & Gao, C. (2020). A Review on Graphene Fibers: Expectations, Advances, and Prospects. Advanced Materials, 32(37), 1902664. https://doi.org/10.1002/adma.201902664